THE ASTRONOMICAL ORIGIN OF CHESS
—the true story—

"The Astronomical Origin of Chess, the True Story"

This fascinating compendium merges two revealing investigations that explore the intricate connections between chess, the Golden Ratio, and medieval hidden cosmic knowledge. By examining the origin of chess, a historical link is revealed with information prohibited by the Church in the Middle Ages, exemplified in tragic episodes such as those of Giordano Bruno and Galileo Galilei. Bruno, consumed by flames, and Galileo, retracting to save his life, were victims for challenging the geocentric belief.

In the first investigation, "Chess Reform and the Golden Ratio," the mysteries behind the structure of modern chess are unraveled, revealing how this strategic masterpiece is based on the mathematical harmony of the Golden Ratio.

The second investigation, "Chess Reform and the Secret of the Cosmos," immerses us in the intricate fabric of medieval knowledge, where Spanish Jewish sages defied the Church's prohibitions to preserve and encode cosmic wisdom. From predicting a Planet X to connecting with the Golden Ratio, this work redefines our understanding of the cosmos and unravels the secrets that official history has attempted to conceal.

"The Astronomical Origin of Chess" is an extraordinary journey through time and space, where the millennia-old game and cosmic mystery converge in a mathematical symphony that challenges expectations and invites us to question established narratives. Delve into this unique compendium that unites the art of chess with the mathematical and cosmic wonders that have remained hidden for centuries.

Tabla de contenido

CHESS REFORM AND THE GOLDEN NUMBER 4
- Abstract 5
- Introduction 6
- The Golden Number 9
- An astronomical approach 13
- The Golden Number in chess 17
  - Pawn promotion 21
  - The double step and the en passant capture 24
  - Fibonacci 25
- Conclusions 31
- Bibliography 32
- APPENDIX 1 33

Chess reform and the secret of the cosmos 36
- **Abstract** 37
- **Historical context** 38
- **The Chess Game of Love** 38
- **The end of the skein** 41
- **The proposal** 43
- **Conclusions:** 47
- References 48

CHESS REFORM AND THE GOLDEN NUMBER

Juan Reyes La Rosa

CHESS REFORM AND THE GOLDEN NUMBER

Juan Reyes La Rosa[1]

Abstract

The present work attempts to demonstrate that the total mobility of the pieces in the game of chess is in exact relation to the Fibonacci sequence. In order to do so it has been necessary to prove that the mobility of the pawn, advancing from the first towards the eighth rank of the board, or vice versa, is 112 and not 140, as it is believed.

Taking into consideration the new value for the total mobility of the pawn, it becomes possible to show that the series, 112, 112, 224, 336, 560, 896 and 1456, which corresponds to the total mobility for each one of the pieces, is obtained by additions beginning from the number 112, which is the value of the total mobility for the pawn advancing in one direction at the board. If these values are divided by 112, we obtain the original Fibonacci series, 1,1,2,3,5,8,13.

The findings of this work allow us to conclude that the game of chess has been reformed so as to achieve the harmony and beauty of the Golden Number or Divine Proportion and that, therefore, its reforms throughout history have a strictly mathematical basis.

Keywords: Chess, Golden Number, chess reform, Fibonacci.

---

[1]Business Administrator and Public Accountant. Correspondence Chess National Champion of Peru, and Candidate FIDE Master.

Introduction

Chess is a board game whose origin has not been attributed with certainty to any of the towns where it was practiced. Neither is there a definitive position about its creators, since several authors, since the Middle Ages, have defended that its creation was performed by Gods, hierarchs and even simple shepherds. What is true is that the way of playing it has mutated throughout time achieving the way and method of playing it nowadays.

What the game represents is another of the great enigmas. In the Middle Ages it was believed that it represented a social structure, with the pawns representing the people and the major pieces, the nobility. It has also been thought to be a bellicose confrontation because the chess was composed of elephants and war carts as combat elements.

Recent investigations assure that everything changed when at the end of the Middle Ages the game was reformed in what respects to some of its pieces' movements, names and forms, gaining since then a harmony never known before in it. Much has been written about this reform and it has given rise to many theories which attempt to explain, for instance, the reasons behind the appearance of the bishop and queen, and why these pieces have the number of moves they enjoy today.

At this point, from the study of the poem Scachs d' amor[2], dated in 1475 and composed by the Valencian poets Castellví, Vinyoles and Fenollar, it has been postulated that the movement of the new queen is due to the political figure of Isabella the Catholic, Queen of Spain,

---

[2] https://es.wikisource.org/wiki/Scachs_d%27amor

due to her outstanding role in the leadership of the kingdom in contrast to the more modest role of Ferdinand II, her husband[3].

However, considering the full title of the poem we note that it is related to astronomical themes, as the title reads:

"*Hobra intitulada Scachs d'Amor, feta per don Franci de Castelvi e Narcis Vinyoles e Mossen Fenollar, sots nom de <u>tres planetas, ço es Març, Venus e Mercuri, per conjunccio e influencia dels quals fon inventada.</u>*"

Manifestly and unequivocally (the underlining is mine) the title tells us that the work 'The Chess Game of Love' ('Scachs d' amor'), was invented by conjunction and influence of the mentioned planets. The work is a poem describing the moves of a chess game and, considering that it contains new pieces and moves, it could be deduced that these were invented by conjunction of the referred planets.

What is new is the name of the apparent new piece and its moves. It is the lady as an evolution of an already existing piece, the primal queen[4].

Stanza I mentions the primal queen, but the successive stanzas, when referring to *the same piece*, call it the lady. This seems a deliberate act so that there is no doubt as to which piece is being replaced. It could even be said that from the poem onwards that piece was no longer called queen but lady (which gives rise to the chess of the lady as opposed to the old chess). The poem makes 8 references to the piece as the queen and 47 as lady, the latter being the star of the game by giving checkmate to the opposing king.

If we look at the details, we will see that the moves of this piece seem to allude to the title of the poem 'The Chess Game of Love' ('Scachs d' amor'), that is also literally translated as 'Checks of Love', since in the game it is the lady who inflicts, of her 5 moves, 3 checks, the last one being checkmate. And, considering that the triumph corresponds to the red side of Mars, conducted by the poet Castellví, a side named

---

[3] Gobert Westerveld postulates Queen Isabella in his book 'Queen Isabella The Catholic: its reflection in the powerful lady of Valencia, cradle of modern chess and the game of checkers'.

[4] It is curious that, existing a primal female piece called queen, it went through being called lady. The explanation of this curious occurrence will be the subject of a future article.

'Amor', it is understood in a metaphorical sense that the green king dies due to the checks of love[5].

The present work contributes a totally different point of view by pretending to demonstrate that the reform was simply due to mathematical reasons; particularly because of the construction of the game in harmony with the Golden Number[6].

Taking as premises that the total mobility value of the rook[7] is the *sum* of the mobility values of its preceding pieces in importance, the bishop and the knight; and, that the mobility of the queen is the *sum* of the mobility values of its preceding pieces, the rook and the bishop; I propose the hypothesis that the mobility value of the pieces in chess corresponds to a pattern of the Fibonacci series, since the values of this series are obtained by the sum of its two preceding ones.

For the hypothesis to be corroborated, the mobility of the pawn must be equal to the difference between the mobility of the bishop and the knight; that is: $560 - 336 = 224$. However, the analysis led me to discover the value of a piece that was supposed to be forgotten with the appearance of the lady: the primal queen[8]. The analysis carried out proves that, indeed, the mobility of the primal queen reached the value of 224, and that the pawn, advancing in only one direction, is 112, thus fulfilling that the mobility of all the pieces is the sum of the mobilities of the preceding pieces.

---

[5] However, what is significant here, veiled if you will, is that the checkmate is inflicted on the green king -since they play red against green-, linking this curious circumstance with the narrative of the church of Rome, imposed in the Middle Ages (which many scholars have denounced), of the stories of pagan gods of vegetation. This would also justify the poets' desire to label both sides with Love and Glory when, apparently, it was unnecessary since one was Mars, the other Venus; one was red, the other green; and one conducted Castellví while the other, Vinyoles.

[6] For more information of this ratio you can visit:
https://www.bbvaopenmind.com/ciencia/matematicas/fibonacci-y-la-proporcion-aurea-geometria-divina/

[7] The total mobility of each piece is obtained on an empty board by counting the squares it dominates when it is placed on each of the 64 squares of the board. See annex 1.

[8] Original queen whose moves were limited to one square diagonally.

The Golden Number

'Geometry has two great treasures: one is the Theorem of Pythagoras; the other, the Golden Number. The first we may compare to a measure of gold; the second we may name a precious jewel'.

Johannes Kepler

In art and nature

The Golden Number[9], or Golden Ratio, is an irrational number with many interesting properties. In ancient times it was also known as divine geometry because it reproduced the behavior of nature, such as the distribution of the leaves of the trees, the thickness of branches, the shell of a snail, etc. The Greeks associated this number with beauty. Many researchers maintain that this golden ratio is present in the most important pictorial works such as The Mona Lisa, by the brilliant Leonardo Da Vinci; Las Meninas, by Diego de Velásquez; Adam and Eve, by Albrecht Dürer; among others.

In the human body, the distance from the navel to the soles of the feet is golden with respect to the total size of the person. The distance from the shoulder to the fingertips is golden with respect to the distance from the elbow to the fingertips, and the latter is golden in relation to the forearm. In addition, hurricanes and spiral galaxies tend to form a golden spiral.

The golden ratio is obtained, using the values of the Fibonacci sequence, by dividing the largest number by its precedent; each

---

[9]For more information about the Golden Number and its application in different aspects of nature and the cosmos, you can visit: https://www.goldennumber.net/solar-system/

number in the series being, at the same time, the sum of its immediate precedents. When dividing a number by its precedent, the quotient approaches in an oscillating way to the value 1.618033988..., the Golden Number or Divine Proportion. Let's look at the following table for a better understanding:

Sequence	1	1	2	3	5	8	13	Fibonacci sequence
Ratio		1.000	2.000	1.500	1.667	1.600	1.625	Golden Number

Illustration 1. Fibonacci sequence numbers and the ratio, the Golden Number

Here it is proved that, 1+1=2; 2+1=3; 3+2=5; ... 21+13=34. At the same time, it is also true that: 1/1=1; 2/1=2; 3/2=1.5; 5/3=1.667; ... 34/21=1.619, tend to the Golden Number.

Phidias (480 B.C.) was a Greek sculptor and mathematician who studied and applied the golden section in his works. He is credited with the conception of the Parthenon temple, whose construction shows the application of the golden section, later defined by Euclid (365 B.C.) in his book *'Elements of Geometry'*.

Although the Divine Proportion or Golden Number has a geometric aspect, it is also possible to study it from the perspective of arithmetic, as occurred with the discovery of the Fibonacci sequence.

Fibonacci, actually Leonardo Bigollo or de Pisa, was an Italian mathematician born in 1175 and died in 1240. He is said to have pioneered the introduction of the Arabic number system in Europe. He presented his famous series studying the way rabbits reproduce arithmetically, proving that, as time went by, the pairs of rabbits were by month: 1, 1, 2, 3, 5, 8, ... 89, etc.

Now, with the series 1, 1, 2, 3, 5, 8, 13, ..., it is possible to form the golden rectangle, considering these numbers as the value of the side of a square. Everything starts with a square of side 1, to which is attached another square of side 1 also, forming a rectangle of side 2; then a square of side 2 is attached so that a rectangle of side 3 is formed. To this rectangle is added a square of side 3 forming a new rectangle of side 5; and so on, successively, by adding a new square we obtain a new rectangle whose proportion of the sides is unfailingly close to the Golden Number 1.618033988...

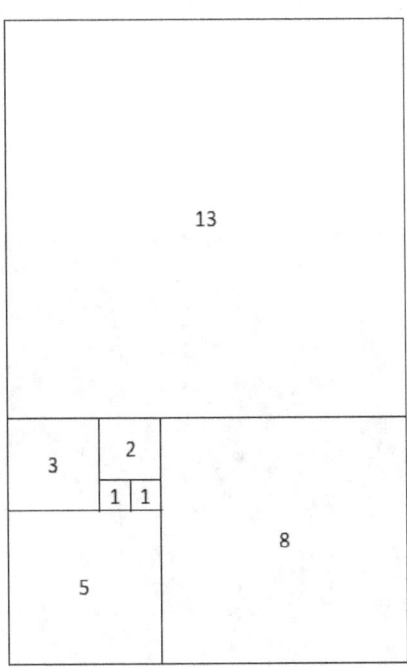

Illustration 2. Golden rectangle of sides 21 and 13 formed by addition of squares

Throughout our existence we live with this golden rectangle: the dimensions of our ID card, driver's license, cinema and television screens, notebooks, wallets, calculators, laptops, etc. Furthermore, it is curious that, by removing a portion in the form of a square from the rectangle, the remaining rectangle is also golden.

And the golden spiral is obtained by joining the vertices of each square, as shown in the graph below:

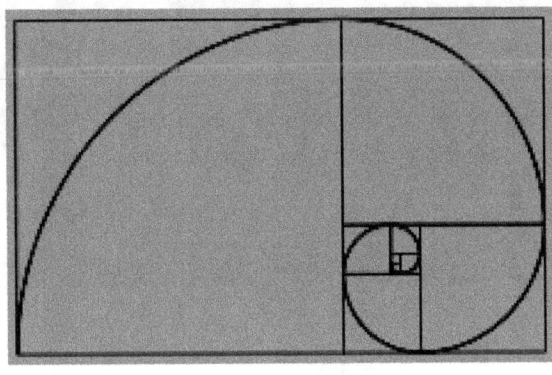

Illustration 3. Golden spiral

As commented before, many phenomena of nature develop following the shape of this spiral, as we will see in the next images:

Illustration 4. Galaxy formation follows the golden spiral

https://hipertextual.com/2015/08/numero-de-oro

Illustration 5. Golden spiral in nature

https://hipertextual.com/2015/08/numero-de-oro

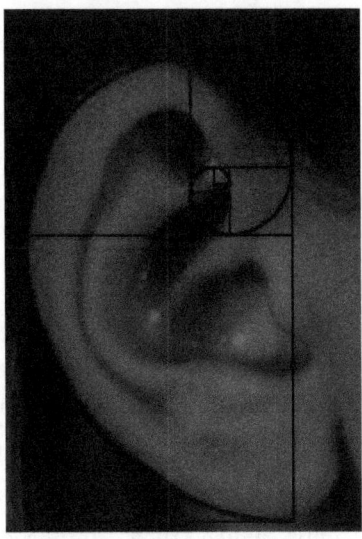

Illustration 6. The shape of the ear follows the golden rectangle.
https://hipertextual.com/2015/08/numero-de-oro

An astronomical approach

Before continuing, it is worth noting that the Valencian poets who composed 'Scachs d'amor' were Jewish converts. However, many authors treat their research as if the word "convert" were magic and prevented the Jew from returning to his diverse activities; nothing could be further from the truth. The truth is that, faced with the atrocities of the armed wing of the Church of Rome, the Inquisition, many converted only by word of mouth, just as some church authorities, in the face of such injustices, supported the cause of the Jews. *'They [converts] were aware of the danger of being denounced by a friend or neighbor, but still, in most cases, they remained loyal to their former religion'*[10].

Therefore, considering the context of suffocating religious repression in the Middle Ages, it is possible that the poem is more an expression of protest or denunciation, if not a version of another reality, opposed to the discourse established by the cultural imposition of the time. It was the Jews, with their tenacious search for truth, who were the protagonists of the Valencian Golden Age. They made business, art

---

[10]Peral Juárez, María Teresa. «The social biography of a Valencian Jewish convert at the end of the 15th century: Úrsula Amorosa.». Annals of the University of Alicante. Medieval History, N. 21 (2019-2020): 115-144, DOI:10.14198/medieval.2019-2020.21.05

and knowledge flourish. For that reason, it would not be surprising that they had astronomical knowledge, as it seems to contain the 'Scachs d' Amor' poem, when its authors declare that it was invented by conjunction of the planets Mercury, Venus and Mars and, obviously, the Earth.

The conjunction referred to in the poem seems to refer to the fact of aligning them for academic purposes, in a map or mockup, in order to show their distances, ratios and proportions with respect to the Sun as the center of the Galaxy[11]. And they would have proposed it applying Hermes Trismegistus' principle: 'As above, so is below'. The mathematical ratio with which they supposed the universe was ordered, should be found in the same way in the order of nature on Earth[12].

This seems to be the reason for the poem to be declamated in the presence of four elements: Mercury, represented by Bernart Fenollar; Venus, by Narcis Vinyoles; Mars, by Francesch Castellví; and, the Earth, represented by the chessboard. In any case, the poem seems to suggest that the board is the Earth since "above", in the cosmos, our planet is placed between Venus and Mars as, "below", at the moment of the poem's declamation; the board is in the middle of the players whom represent both planets.

That the Earth was not the center of the universe has been known since Aristarchus of Samos, 300 years before our era. The same happens with the distances from the planets to the Sun, knowledge that was made public with the work of Nicolaus Copernicus in 1543,

---

[11] "Although in the third century B.C. Aristarchus of Samos had already pointed out that the Earth and the planets revolved around the Sun, and although Archimedes echoed this, the prevailing theory was the geocentric theory described by Ptolemy in the second century" ('Scientific Thought' journal, Polytechnic University of Madrid, Volume IX, No. 2).

[12] As Ibn Ezra (1089-1164) also believed: 'the order and superior stellar movements represented the order of God and, therefore, govern human events'. (Ana R. Gonzáles Sánchez. Doctoral thesis: 'Tradition and fortune of the astromagic books of the Alphonsine scriptorium'. P. 241.) Alfonso X also believed: 'the function of the wise astrologer/magician is to acquire the maximum knowledge of the stars, in such a way that this knowledge allows him to modify the influence of the celestial beings on the terrestrial ones. This is the magic that, as we will see later, the Wise King professed'. (H. Salvador Martínez. 'Alfonso X the Wise, humanist and scientist'. Argutorio Journal 40, 2018.

at his death. It is possible that the chess reformers, of the same school of the Valencian poets, may have considered the following values of distance towards the Sun[13] and recognized the Golden Number:

Mercury	0.386
Venus	0.719
Earth	1.000
Mars	1.520

These values are expressed in relation to the distance from the Earth to the Sun; that is, taking the distance from the Earth to the Sun as a unit. Now, if we consider the planets enunciated by the poets, we obtain the following value:

Mercury		0.386
Venus		0.719
Mars		1.520
	Σ	**2.625**

And it happens that this number, 2.625, is related to the Golden Number in the following way: 1 + 1.625 (Approximation to the Golden Number), and its graph would be as follows:

```
      1            1.625
 ┌─────────┬──────────────────┐
 │         │                  │
 │         │                  │
 └─────────┴──────────────────┘
```

In superimposed planes it would look like this:

```
        1 + 0.625
   ┌─────────┬────┐
   │         │    │
   │   1     │    │
   │         │    │
   └─────────┴────┘
```

This rectangle is golden because the ratio of its sides is 1.625. And it might have been used to relate their astronomical observations to this figure's interesting geometrical properties, since the distance of Mercury to the Sun approaches this value: (1/1.625)-1= 0.384. Or,

---

[13] According to Copernican calculations

perhaps, in this way: 0.625 x 0.625 = 0.390. Earth would be 1 x 1, Mars would be 1 x 1.625 at some point in its elliptical rotation.

	1 + 0.625	
1	0.39	.6 2 5

It is not known if they suspected the existence of the imaginary ghost planet Vulcan, in force until the 20th century, making the calculation 0.375 x 0.375 = 0.1406, because in 1859 the French astronomer, Le Verrier, proposed its existence and estimated an average distance to the Sun of 0.1427[14].

	1 + 0.625		
1	0.39		0.625
		0.14	0.375
		0.375	

What is really important in this analysis is that it results in the application of the Golden Number to explain the harmonic functioning of the planets and, of course, it seems that this information was known in a reserved way among astronomers of that time, mostly Jews, who since the twelfth century would have been grouped in Schools of Translators of all human knowledge[15]. However, of all these calculations, we will pay special attention to the number 1,625 that would appear with the reform of the game of chess, as we will see in the following section.

---

[14] https://www.taringa.net/+ciencia_educacion/cuerpos-hipoteticos-del-sistema-solar-v_137ioi

[15] "The most important contribution of the School, in any case, is that related to scientific disciplines: medicine, mathematics, astronomy and astrology. The aforementioned canon of Avicenna, Islamic trigonometry, the sexagesimal system, the Astronomical Tables of Al-Khwarizmi and the works sponsored by Alfonso X, would be the starting point of European astronomy and the reference sources for the works of Galileo, Copernicus, Kepler or Newton." Ana R. Gonzáles Sánchez, op. cit. P. 18.

The Golden Number in chess

'Chapter one. Game of chess, game of science and mathematical invention' (López de Segura, 1561, p. Fol. 1).

Svetozar Gligoric's hint

I had not noticed the existence of this number until I wrote my first book. It was very curious that many figures related to the golden spiral adorned Templar churches; particularly the nautiluses. My curiosity was even greater when I found out that the Templars built their churches with an octagonal base and had special consideration for the eight-pointed cross, like the Maltese Cross. To add to my suspicion, the Masons, supposed heirs of the Templar secrets, insisted on the number eight by including in their symbology a chessboard with eight squares on each side.

In the midst of my musings I remembered reading a book based on the master classes offered by GM Svetozar Gligoric[16] to the Spanish chess team, back in 1985. Gligoric referred that in chess there is a relation between the value of the pieces and a number like 1.61. That number was like a thunderbolt that made me jump from the armchair. He refers:

---

[16]Svetozar Gligorić, was a renowned Yugoslav International Grandmaster chess player. In the 1950s and 1960s he was one of the strongest players in the world, standing out also as a theorist, commentator, pedagogue and disseminator of the game. Wikipedia. Probably, American genius and world champion, Robert Fischer's closest friend.

'Various scholars on the subject have found an almost exact correlation between values of the diverse chess pieces. Thus, the relationship between the pawn and the knight would be 1 to 1.61; and likewise between the bishop and the rook or the rook and the queen.

It seems, then, that there is an exact mathematical relationship between the strength of the pieces, measured according to their efficiency. This argument suggests that there is a very deep conception of how the pieces should move.

Of course, in a chess game, these values are variable, since they also depend on the activity of the other pieces and the structure of the position. We already know that, when the game is closed, the knight is worth more than the bishop and the bishop is more powerful than the knight when the position is open. But, if we look at the efficiency of the bishop and knight on an empty board, when moving from h1 to a8 and when calculating how many squares these pieces dominate, the mathematical relationship between both will be 1 to 1.61.' (Gligoric, 1985)

Although not mentioned by master Gligoric, I was almost convinced that this mysterious number had something to do with the Fibonacci series and, thus, with the Golden Number. My curiosity reached its highest level when I began to deduce that the chess reform would have geometrical and arithmetical motifs related to the harmonic beauty that I already explained and that, therefore, it had little to do with troubadours and poets whom they have pretended to relate to the reform of the game in the 15th century[17].

The first book that aroused my suspicion was '*Mitología del ajedrez*', by Francesc Cardona. He wrote:

---

[17] 'The stilling proof of the Spanish origin of modern chess is the poem Scachs d' amor (Valencia, c. 1475), where the chess of the lady is given naturalization certificate, and in unequivocal terms that it is they, the poets Fenollar, Vinyoles and Castellví, who have created the new piece, the lady, and the rules that derive from that revolution, which is maintained today, in full technological age of chess'. (Garzón, eHumanista 47, 2021, p. 215)

'There are **728** different ways to place two queens of the same color so that they do not protect each other. For rooks, the number is **448**. For two bishops, **280**. For two knights, **168**.' (Cardona, 2000, p. 113)

Then I took my calculator and added: 168+280=**448**; 280+448=**728**. Fibonacci! I wrote in red ink in the book: the higher responds to the sum of its precedents.

Then I dusted off my Chess and Math book and was able to find the following:

- 'Without taking castling into account the king can execute 420 different moves on the normal board, [...] and the other pieces execute: Q= 1456; R= 896; B= (of white or black squares) 560; N=336; and P=140; in this, the pawn promotions are not separately valued'. (Bonsdorff, Fabel, & Riihimaa, 1974, p. 56)

Again, calculator in hand, I added: 336 + 560 = **896**; 560 + 896 = **1456**. Fibonacci! I wrote in red again and marked the quote from the book.

Then I looked up the quotients:

560/336 = 1.667.

896/560 = 1.600

1456/896 = **1.625**

These numbers were similar to the ones explained by master Gligoric.

Nevertheless, 140, which is the mobility of the pawn, does not fit the series. For the pawn to be part of that beautifulness that is achieved with the Golden Number, the value of its mobility should be: 560 - 336 = 224. Or, what is the same, 112 + 112, because, considering that the pawns cannot retreat, to determine its mobility we must estimate it separately considering white and black[18].

Let's return to the subject of the mobility of the pawn, which was said to be 140, and let's see how it would have been valued. For its

---

[18]The same is not true for the other pieces since, in order to determine their total mobility, color is disregarded as they move forward and backward.

calculation, the squares that dominate the set of pawns from its initial position to the seventh rank, would have been added.

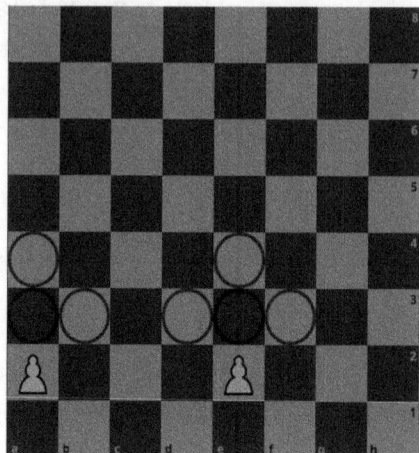

In the graph above we observe how many squares the pawns dominate in their initial position, namely: the rook pawns, 3 squares; the central pawns, 4 squares. Considering 2 rook pawns and 6 center pawns, they add up to 30 squares from their initial position. Now let's see when they advance one square:

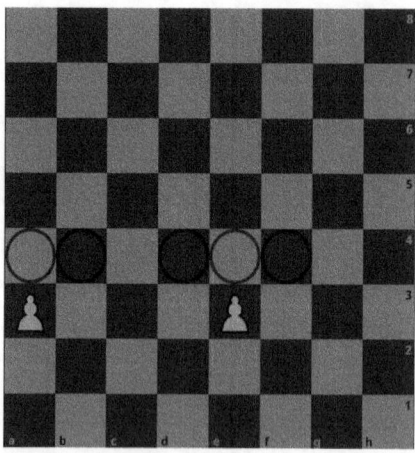

Here we see that the rook pawn dominates 2 squares, while a central pawn dominates 3. Considering 2 rook pawns and 6 central pawns, they add up to 22 squares which, multiplied by 5 ranks (from the third to the seventh), add up to 110 squares.

In summary:

Posicion of the pawn	Dominates squares
In their inicial posicion	30
From the 3rd to the 7th rank	110
Total	**140**

The problem seems to be that this value does not take into consideration their special movements, such as promotion and en passant capture[19].

a. **Promotion:** Consists in replacing the pawn when it reaches the last square, by a queen, rook, bishop or knight.

b. **The double step and en passant capture:** the first one consists in the player's posibility to decide whether to advance his pawn, when it is in the initial position, two squares, according to what he sees appropiate in the position. However, this double step has, during the opponent's immediate turn, an associated posible move, the en passant capture, which *reduces the mobility* of the opponent's pawn, since this move can convert the double step into a single one, by capturing it one square behind where it is placed after its double step.

Pawn promotion

To say something concrete about the promotion of the pawn, it is better to say it in the words of the great José Raúl Capablanca:

'There are a multitude of details in the history of chess that demonstrate in a clear way to have been subject to a continuous evolution through the centuries, and probably will experience more evolutions in the future. In fact, barely a hundred years ago, it was still a matter of discussion whether a pawn on reaching the eighth square could be transformed into the piece of the player's choice '(J. Ganzo, p. 19).

---

[19]This is recognized by the authors of *'Chess and Mathematics'*, Bonsdorff, Fabel, & Riihimaa, p. 56.

Experts on the subject agree that, since its creation, several versions of the same game coexisted. The last ones, of course, correspond to the chess of the *'Alferza'* and the chess of the lady or *'a la rabiosa'*; and even this is in doubt since the discovery of *'Scachs d' amor'* poem, because there the predecessor of the lady is punctually and literally the queen.

Unfortunately the poem, which reproduces a modern game, does not contain a pawn promotion move[20]. It seems natural to all of us that the pawn promotes with a capture in the eighth rank.

If we review the original literature referring to the reforms, especially that of the persecuted Jews, we will find, for instance, that the promotion should have converted the pawn only into a queen and not into another piece. What could have happened then? Probably practicality won out in the absence of spare queens, or perhaps the persecutor erased the traces of the reform[21].

The same thing could have happened with the promotion moves, since the literature of the time does not refer to the pawn capturing in the eighth rank. Following Ruy Lopez, we can see that the coronation or promotion of the pawn occurs by its natural gait, when he says: 'they cannot *walk* if not from house to house, until they reach the last of the enemy. Because once they arrive there, because of their virtue they acquire the power and name of ladys' (J. Garzón. p. 54). One might ask, what is the natural gait of the pawn? We will answer, forward, house to house, which could be interpreted to mean that it is only crowned with its natural gait. Lucena, prior to Ruy Lopez, says: 'that when it arrives to its opponent's king rank, it adquires the strength of a lady and gives check without transposing'. And coming to the book of Alfonso X, this wise king says that 'the pawn can be made alfferza after it *walks* six times from house to house'. Everything seems to

---

[20]Nevertheless, his stanza 57 disagrees with the promotion to lady: 'If the ancients, to increase their brood, with no regard to law or justice, from lowly blood and ordinary paste, strive to create a thousand queens out of malice, the laws of amorous malice say that the diamond more lawfully can be mounted and shine with great clarity. The faithful lover falls but for one; the ungrateful infidel does adore the idols'. (en.wikisource.org/wiki/Scachs_d%27love)

[21]The Church has not been on good terms with chess for many years, even going so far as to ban it.

indicate that the promotion could only occur with its natural displacement movement on the board, without considering its capture movement. We have to remember that only the pawn has separate displacement and capture movements. With respect to the other pieces, their advance and capture are part of the same move.

I mean, that the long way of perfecting the game, probably initiated in Castile, is based on harmonizing the movements of the pieces until reaching the beauty of today's game, because its wise reformers, astronomers and mathematicians of Jewish roots, considered pertinent to apply to the mobility of the pieces the harmony that they observed in the universe.

However, if we stick to the text of Ruy Lopez de Segura who states, in his book of 1561, that chess is 'a game of science and mathematical invention'; and, if we consider that the authors of the referred poem state that modern chess 'was invented by conjunction and influence of the planets Mars, Mercury and Venus', conjunction that has a mathematical expression; and if, in addition, we relate these expressions with the explanation of GM S. Gligoric, who assures that the relation of mobility of the pieces, measured on an empty board, is 1.61; I could assure that this mathematical invention and conjunction of the planets are related to the beauty of the harmony that the Golden Number or Divine Proportion gathers.

It seems, then, that because of these considerations the relation between the mobility of the queen and the rook is 1,625, when we divide 1456; the mobility of the queen, divided by 896, that of the rook. And, if we do the same between the mobility of the rook and the bishop, we obtain the value of 1.60. This would reinforce the idea that the mobility of the pawn is, indeed, 112 and not 140. Something happened in the long road of the reform that corrupted its value.

If our interpretation of the pristine rules were correct, we would have to reduce the total mobility of the pawn by 14 capture moves in the eighth rank: 2, for the captures of the rook pawn, and 12 for the captures of the 6 central pawns.

Moreover, it is pertinent to note that the pawns in the seventh rank have 22 promotion moves towards the eighth rank, considering that only 8 promotions can occur, the rest constitute fictitious promotions, which are 14, which reinforces the idea of reducing the total mobility of the pawn by 14 units.

The double step and the en passant capture

The other issue is the en passant capture: a white pawn placed on the fifth rank can convert the opposing pawn's double advance into a single advance by capturing it as if it had only taken a single step, *thus reducing its total mobility*. In this way, the white rook pawns, on the fifth rank, subtract one move each -when the black pawn advances two steps from b7 to b5 or from g7 to g5 and is captured as if it had advanced only one, to b6 or g6-; that is, it subtracts 2 moves from the total. The other 6 central pawns can subtract, each one, two moves; that is, 12 in total, which, added to the 2 of the rook pawns, add up to 14, as shown in the following diagrams.

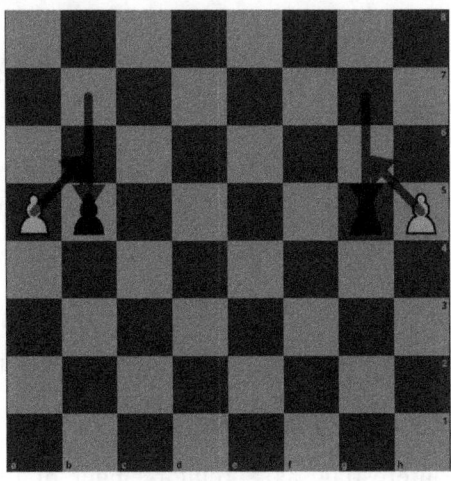

Illustration 7. Black b pawn's double step, loses one square when it is taken by the en passant capture.

Illustration 8. Black pawns, d and f, lose one square due to the en passant capture.

In the diagram, the black pawn loses the position reached with its double step on d5 because it is captured on d6. The same happens with the double step of the f-pawn: it loses the position reached on f5. Note that there can be an en passant capture only if there is a double step of an adjacent opponent's pawn on the board. While the double step adds squares to the total mobility of the pawn, the opponent's en passant capture takes squares away from it.

Fibonacci

With this new data, the total mobility of the pawn, when marching in the direction from 1 to 8, or the other way around, is reduced to 112 as shown in the following summary:

Caurrent movility	140
Adjustments:	
Promotion	-14
En passant capture	-14
New mobility	**112**

This seems to be the sought value since it harmonizes with the value of 224 which is the difference of the mobility of the bishop minus that of the knight, in this manner: 560 - 336 = **224**; 112 + 112 = **224**.

Here it is very interesting, as a reinforcement of this proposal, to include in the analysis the mutation of the queen into a lady, if we believe that the last reform, regarding the mobility of the pieces, is contained in the 'Scachs d' amor' poem. For this, we could assume that the Alfferez, referred to by Alfonso X, is the Pherez of the poem of the Jew Ibn Ezra because, according to Ernesto Negri, in an article of his in Chessbase he says that 'It is curious, although entirely improper, since it does not respect the original spirit of Ibn Ezra that, when Hollaenderski[22] refers to this piece, he "clarifies" that the pherz, so he calls it, is equivalent to the reine, an association that will only be admissible in times to come[23]'.

This idea takes shape when we verify that these pieces preceding the lady: Alfferza, Pherez or Pherz, or primal queen, had, at the beginning of the game, moves limited to one square diagonally; the promoted pawns took their names and in all cases the promoted pawn moved everywhere like the current queen[24]. It seems, then, that the same

---

[22]Jewish author, translator and poet.
https://en.wikipedia.org/wiki/L%C3%A9on_Hollaenderski
[23]https://es.chessbase.com/post/ibn-ezra-articulo-2020-por-sergio-ernesto-negri
[24]And it is even possible that she had full movements from the beginning of the game, thus: 'The rabbi Bonsenior Josehp ben Jachia from Barcelona, who in the middle of the XIII century wrote a sentence about chess, says that in it there is a woman who dominates everything' (J. Brunet and Bellet. 'Chess: Investigations of its Origins' p. 323). According to José Rodríguez de Castro, interpreting Ezra's poem, he says: 'the Queen, whose office is to guard her lord, goes wherever she wants, because she has for hers all the roads (...) Of all the King's army the Queen is the foremost person and she has more power and faculties to defend him than the others'. (J. Brunet and Bellet, op. cit. p. 358).

piece had two distinct moves; restricted at the beginning of the game, and free when it comes from a promotion.

Ibn Ezra, when referring to the Pherez' movements, says in his poem that: 'When the Pherez walks it has an entrance / In four parts marked'. This verse seems to refer to the restricted movements at the beginning of the game, since the four inmediate squares diagonally from the position it occupies are the greatest mobility of this piece. Nevertheless, when referring to the promoted pawn, it moves as a distinguished Pherez, with full reach to the whole board according to his poem: 'If by good fortune walking/ much from its place it moves away/ and up to the eighth order it has climbed/ as if it were a distinguished Pherez[25]/ it can then become/ and free everywhere resolve itself'[26].

Alfonso X expresses this about the promoted pawn: 'it be alfferzed as many times, as the other alfferza, walking all the houses of the board, which it can walk[27]'.

It seems, then, that with the last reform they stopped using one of the moves of a piece that always existed, the primal queen: its initial movements. They have kept the same piece, giving it a new name, but with the freedom of mobility that always belonged to it; that is to say, the new lady takes the place of the primal queen with the ancient powers of a newly promoted pawn[28]. If this were the case, the reform began in Castile[29], as Ruy Lopez asserted, where Alfonso X -sometime nicknamed the astrologer[30] - reigned, who would have invented the astronomical chess (Martínez Sáez, 2021) to represent the superior order of the cosmos and would have conceived the 8 x 8 chess as an

---

[25] Note that the author only refers to promotion to distinguished Pherez; that is, to queen.
[26] https://es.chessbase.com/post/ibn-ezra-articulo-2020-por-sergio-ernesto-negri
[27] J. Brunet and Bellet. op. cit. p.248.
[28] Perhaps for this reason, the Valencian poets intended to annul the queen promotion since it was already on the board from the beginning of the game.
[29] Ruy Lopez says: 'that, with respect to the inventor of such a noble game, he does not think he has to spend many words to know whether it was the Moors, as some believe, which is not true, because the game was known before they came to Spain [...] -I will not say something else, but that for many reasons it could be proved to have come from Castile, and that it is clearly enough known that its inventor is much to be praised for' (J. Brunet and Bellet. op. cit. p.307).
[30] Ana R. Gonzáles Sánchez. op. cit. P. 18.

earthly representation, —political-military— of the Spanish medieval State (Martínez Sáez, 2021); and where the foundational ideas of the modern queen also come across with the Pherez, of Ibn Ezra, and the Alfferez , of Alfonso X. Ibn Ezra, it is pertinent to mention, had a notable influence on the Castilian Kabbalistic School[31], who was also an astronomer, as Ernesto Negri wrote: 'He was someone who knew how to define the division of the planet into two parts based on the conception of the Equator. A notable Andalusian Jewish intellectual, (...) who would stand out interpreting the tradition of the sacred books, who also made contributions in *astronomy*, philosophy, grammar, science and many other fields of knowledge[32]'.

The curious thing about all this is that the lost moves of the primal queen, whose existence is credited since the twelfth century, or earlier[33], add up to 224. Performing the exercise to determine the total mobility of this piece, traveling the entire empty board with its limited gait, adds up to 196[34]. Let us add her two special movements that are not consistent with the current movement of the queen: 1. The jump of joy; and, 2. The warlord's move. In the first case, the newly promoted queen could, in her first move, jump like a knight[35], which adds to her mobility 26 moves. The warlord's move refers to the ability of this piece, at the beginning of the game, to jump in its initial move over its pawns up to the third square diagonally[36], which adds 2 moves (Qb3 and Qf3) to its mobility. Adding these lost moves we have the following summary:

---

[31] Ana R. Gonzáles Sánchez, op. cit. Pág. 43.
[32] https://es.chessbase.com/post/ibn-ezra-articulo-2020-por-sergio-ernesto-negri
[33] The first mention of a chess piece as a queen dates back to the 10th century, in the 'Versus de Scachi' poem, by an anonymous author. (https://www.youtube.com/watch?v=I6W5wi2Sg5E)
[34] See the appendix that refers to the calculation of the mobility of the pieces.
[35] 'That for the first time that she enters as lady and the first throw that she plays, she gives check as a lady and a knight' (J. Brunet and Bellet. 'Chess: Investigations of its Origins', quoting Lucena. p. 370).
[36] 'Murray points out that the Férez is allowed the extended move on the first move by jumping to the third square while retaining the color of the original square'. https://es.chessbase.com/post/ibn-ezra-articulo-2020-por-sergio-ernesto-negri

Natural movements	196
The jump of joy	26
The warlord's jump	2
lost movements	**224**

In this manner, the Fibonacci pattern is met, which consists of obtaining an ordered series of numbers that respond to the sum of their two preceding ones:

Σ	Mobility	Piece
	112	White pawn
	112	Black pawn
112+112	224	Primal queen (lost)
224+112	336	Horse/Knight
336+224	560	Alfil/Bishop
560+336	896	Rook
896+560	1456	Lady/Current queen

Therefore, the harmonious mobility of the new chess -the chess of the lady that we practice today-, should derive from these values:

112, 112, 224, 336, 560, 896, 1456.

If we divide all by the smallest value of the series, 112, we would have:

1, 1, 2, 3, 5, 8, 13, the Fibonacci sequence.

And, in fact, as I have already explained, this series contains the values to affirm that the moves of the new chess, which they call 'of the lady' -to paraphrase Lucena-, are harmonized with the Golden Number[37], the Divine Proportion, as can be seen in the following table:

---

[37] On February 26, 2008, 'El Universal' reported that architect Franco Rocco came to the conclusion that the engraving of the chess pieces found in Luca Paccioli's *'De ludo Schacorum'* (1500), recently discovered in a private collection, belong to Leonardo Da Vinci himself. Among his arguments is the fact that the golden ratio is contained in the design of the pieces.

Pieza	Σ	Mobility	Ratio	
White pawn		112		
Black pawn		112	1.000	
Primal queen (lost)	112+112	224	2.000	
Horse/Knight	224+112	336	1.500	
Alfil/Bishop	336+224	560	1.667	
Rook	560+336	896	1.600	
Lady/Current queen	896+560	1456	**1.625**	**Golden Number**

For this reason I am inclined to think that the new chess is a deliberate creation; it is not a random occurence in the sense of taking the movements of the pieces as they had arrived in the fifteenth century. I believe, like Brunet and Bellet, that some movements were already in force at the time of Alfonso X, the Wise, as is the case of the rook, the horse - which Sir Alfonso corrects by saying that *their right names are Knights* - and the pawn; They introduced the bishop[38] -nowadays known in Spanish as 'alfil'- and adopted as movements of the new lady, those corresponding to a pawn recently promoted into a queen of the old game, to reach, from the beginning of the game, the beauty of the Golden Number that in the old game of the queen was reached with the promotion of a pawn. The movements of the knight and the rook were not altered because the rook was in golden relation to the knight and the knight was in golden relation to the pawn[39].

---

[38] According to J. Brunet and Bellet, op. cit. P. 332, 'the first who in France gave the Bishop the name of Fal or Fau —madman— was the author of the Roman de la Rose'. Probably Guillaume de Lorris, who introduces in his story the figures of Hermaphroditus and Atis, a Roman cult divinity linked to madness.

[39] There are clues that suggest that the double step and the en passant capture existed during the times of Sir Alfonso, the Wise. Thus, for instance: 'the pawns do not go more than to one house in their right, like the laborers of the army; [...] But there are some who use to play the pawns to the third house when moved for the first time, and this is until they take a house; after that they cannot do it'. (J. Brunet and Bellet, op. cit. p. 247).

Conclusions

1. The reforms of the game of chess have pursued the beauty and harmony that is achieved when its elements reach the Golden Number.

2. The mobility value of the pieces are in exact relation with the first values of the Fibonacci sequence, 1, 1, 2, 3, 5, 8, 13, which is obtained by dividing the total mobilities of each piece by the value of the total mobility of the pawn, which is 112.

3. The game has been reformed on the basis of strictly mathematical considerations as Ruy Lopez assured.

4. GM Gligoric was right: there is a relation in the game of chess expressed in the Golden Number, 1.618033988...

Bibliography

Bonsdorff, A., Fabel, K., & Riihimaa, O. (1974). *Ajedrez y Matemáticas (Chess and Mathematics).* Barcelona: Martínez Roca.

Brunet y Bellet, J. (1890). *El Ajedrez: Investigaciones sobre su Origen (Chess: Investigations of its Origins).* Barcelona: Hispano Europea.

Cardona, F. (2000). *Mitología del Ajedrez (Chess Mythology).* Barcelona: Edicomunicación S. A.

Philidor, A. D. (1887, originally published in 1777). *Análisis del juego de Ajedrez (Analysis of the Game of Chess).* Paris: Librería de Ch. Bouret.

Garzón, J. A. (2001). *En Pos del Incunable Perdido. Francesch Vincent (In Search of the Lost Incunabulum. Francesch Vincent).* Valencia: Biblioteca Valenciana.

Gligoric, S. (1985). *Curso de Entrenamiento (Training Course).* Alicante.

López de Segura, R. (1561). *Libro de la Invención Liberal y Arte del Juego de Ajedrez (The Art of the Game of Chess)* Alcala.

https://web.archive.org/web/20061030223236/http://luppas.homeip.net/astrotaller/viejos_nuevos_planetas/vulcano/vulcano_astronomia.htm

Universidad Politécnica de Madrid. Research journal "Pensamiento Matemático", volume IX, number 2: Aspectos científicos del viaje del Descubrimiento (Scientific aspects of the voyage of Discovery).

Ana R. Gonzáles Sánchez. Tesis doctoral: Tradición y fortuna de los libros de astromagia del scriptorium alfonsí (Doctoral thesis: Tradition and fortune of the astromagic books of the Alphonsine scriptorium). Department of Spanish Philology of the Faculty of Philosophy and Letters of the Universidad Autónoma de Madrid.

H. Salvador Martínez. Alfonso X the Wise, humanist and scientist. Argutorio Magazine 40. 2018.

APPENDIX 1

DETERMINATION OF THE MOBILITY OF THE PIECES

The total mobility of a piece is calculated by adding up the squares that it dominates from its position, going across the 64 squares of the board.

1. Mobility of the current queen/lady

In the diagram above, the current queen or lady, placed on a1, dominates 21 squares, which is equivalent to 21 possible moves. When it is in the center, it dominates 27. If we add up the squares that it dominates in each of its 64 positions, the queen's mobility reaches the value of 1,456.

2.   Mobility of the rook

The total mobility of the rook, as can be seen, is calculated by multiplying 64 x 14, since the rook dominates 14 squares from any placement on the board. Its mobility reaches the value of 896.

The mobilities of the bishop, the knight and the pawn are calculated in the same way.

3.   Mobility of the primal queen

As can be noticed in the diagram, when the primal queen is placed on d1, it dominates two squares; when it is placed on h1, it dominates one square because it is a rook file. Then, considering only the first rank, its total mobility -without the special moves- would be: 2 x 6 central placements = 12; plus, 1 x 2 lateral placements = 2. A total of 14 squares. The same is repeated in the next rank, so it reaches 28 squares.

When the primal queen is placed in any rank, from the second to the seventh, its mobility per rank is: 4 x 6 central placements = 24; plus, 2 x 2 lateral placements = 4. A total of 28 squares per rank that, multiplied by 6 ranks, = 168 squares.

In summary: 28 squares of the extreme ranks, plus 168 squares of the intermediate ranks = 196.

Chess reform and the secret of the cosmos

Chess reform and the secret of the cosmos

Juan Reyes La Rosa[1]

To the courageous souls of medieval Spain who, in the face of insurmountable adversities, dedicated their lives to the unwavering pursuit of truth and freedom of thought.

To Giordano Bruno and all those who defied circumstances to contribute to the advancement of knowledge and the exploration of ideas.

Abstract

This research presents a novel methodology using the Golden Number to estimate planetary distances to the Sun. Through a geometric and quadratic progression based on the distance of Mercury, $M=(\Phi-1)^2$, all planets are encompassed, including the prediction of a possible Planet X. Contrary to scientific expectations, this study suggests that Planet X would lie between Saturn and Uranus, at a distance of approximately $(10M)^2$ AU, challenging current hypotheses about its location beyond Pluto. Furthermore, the progression excludes Ceres, but claims Pluto as a planet.

---

[1] Business Administrator and Public Accountant. Master's studies at the Universidad Nacional Mayor de San Marcos and International Diploma in Management Control at the Universidad de Piura in association with the Universidad de Chile. National Distance Chess Champion of Peru, and Candidate Master by the International Chess Federation.

Historical context

When I finished this work, it seemed to me that a planet was missing. This is what it seems to a layman in astronomy as I am. That's why I don't believe it, especially if the vertiginous progress of science has almost everything watched and explained. And yet, it seems to be missing.

It all started with my suspicions that the written history is not necessarily the true one. I have had my doubts since I understood that history is usually written by the victor. The vanquished may have been right or had noble purposes, but these were buried by the narrative of the facts from the point of view of the one who prevailed. Significant cases can be the facts of the Conquest of America, which resulted in the loss of sovereignty and identity of the native peoples, or the religious persecution that occurred in medieval Europe, which resulted in the stagnation of knowledge and the segregation and elimination of peoples who opposed Catholic orthodoxy.

What has all this to do with the planets? The subject of the planets was another aspect in which medieval religion set its conditions to the point of excommunicating and punishing, if possible with death, those who tried to alter the Ptolemaic order that the church defended. The most emblematic case is that of Galileo Galilei, in 1633, condemned to recant and suffer house arrest until his death for affirming that the planets orbit around the Sun, as postulated by Nicolaus Copernicus. Nicolaus himself waited until his death for his heliocentric theory to be made public to avoid the disgrace of the Inquisition snooping on his work. However, Galileo would have seen himself in the mirror of Giordano Bruno who was burned at the stake, in 1600, for affirming that the Sun was the center, the planets orbiting around it and the infinity of the universe. Bruno refused to recant knowing that his flesh would be burned and his books banned and burned.

The Chess Game of Love

For this reason, any work by an anonymous author or signed with a pseudonym attracts my attention, because it is possible that it contains information that harms someone. This is the case of the 15th century Valencian poem, Scachs d Amor (The Chess Game of Love), discovered by a Jesuit monk in 1905 in the Royal Chapel of Palau, in Barcelona, Spain. This poem reveals for the first time in the history of chess the name of 'dama' (lady) for the most important piece. Although we identify its authors, Vinyoles, Castelví and Fenollar, the title is intriguing:

"Hobra intitulada Scachs d'Amor, feta per don Franci de Castelvi e NarcisVinyoles e Mossen Fenollar, sots nom de tres planetas, ço es Març, Venuse Mercuri, per conjunccio e influencia dels quals fon inventada".

(Work entitled The Chess Game of Love, made by Don Franci de Castelvi, Narcis Vinyoles and Mossen Fenollar, under the name of three planets, that is, Mars, Venus, Mercury, by conjunction and influence of which it was invented).

This is the title for a poem that describes for the first time a chess piece with the name of 'dama' (lady) and, from here, the new chess is known as 'ajedrez de la dama' (chess of the lady), which is the one we practice today. What calls my attention is the willingness of the poets to specify that the poem, which contains moves of modern chess, was invented by the conjunction and influence of the planets Mars, Venus and Mercury. It is understandable that the poets think or dream of the Moon or Venus, but in what way did these planets influence the poem? What relation do these planets have with the very new chess? Why are their authors not explicit?

In a previous article I had demonstrated —I believe— the mathematical (astronomical) origin of chess. I postulated that its construction had been made for mathematical reasons, particularly with the application of the golden number or golden ratio. I had found that the total mobility of the modern chess pieces had an exact relationship with the Fibonacci series. Its famous series, 1, 1, 2, 3, 5, 8, 13, is obtained, in modern chess, by dividing the total mobility value of each piece by the smallest value of the series, that of the pawn which is 112 (Reyes La Rosa, 2022), as can be seen in the following table:

Chess piece	Σ	Mobility	Ratio	
pawn going up		112		
pawn going down		112	1,000	
Primal queen (lost)	112+112	224	2,000	
Caballo/Knight	224+112	336	1,500	
Alfil/Bishop	336+224	560	1,667	
Torre/Rook	560+336	896	1,600	
Dama (lady)/Modern queen	896+560	1456	**1,625**	Golden Number

Here it is true, as in the famous series, that the value of a piece corresponds to the sum of the values of the preceding pieces in importance. Thus, the rook, whose total mobility value is 896, corresponds to the sum of the total mobility of the bishop and the knight. The same with the queen, 1456 is the sum of the rook and the bishop. Now, the quotient of these values, as in the famous Fibonacci series, tends to the golden number, 1.618033989. In the chart, the value of 1.625 is the division of 1456/896, and is the closest approximation to the mysterious number.

Intrigued, moreover, because it is not very widespread or has not been paid (or does not want to be paid) attention to the final part of the title, otherwise cryptic, which says: 'The stanzas are in chain form, with nine lines each and in sequential order, that is, four, three, and two, and thus must they be written and read. In their inscriptions you will see the sum of their literal sense, that is, the game of chess and the pacts to be obeyed'. That is why I set out to find out if the poem contained uncomfortable information for the time, especially since its authors were all Jewish "converts "[2]. Encouraged also because chess and its exponents, at some point in history, were persecuted by the political arm of the church. In fact, the authors of the poem "Scachs d' Amor" did not have a good time after the publication of the poem; it is known that Castellví and Vinyoles were prosecuted by the Inquisition (Pérez Bosch, 2009, p. 54). In addition, Vinyoles had to rescue his wife from the fury of the Holy Office. Let us add to this list Francesch Vicent, who was a Valencian chess player and writer who fled Spain at the end of the 15th century due to the persecution of the Jews (Argos1942, 2020). Vicent was the author of the world's first modern chess treatise, printed in Valencia in 1495, with the title 'Libre dels jochs partits dels schachs en nombre de 100' (Book of the games and games of chess in number of 100). In this book, Vicent echoes the work of his fellow countrymen, the poets who composed "Scachs d' Amor", introducing the modern move of the queen. On the other hand, Luis Ramírez de Lucena, author of 'Repetición de amores y arte del ajedrez', published in 1497, was involved in problems with the Inquisition because they had imprisoned one of his uncles. His own father, Juan Ramírez de Lucena, prothonotary and ambassador of the Catholic kings, lost part of his dignities because of his position against the abuses of the Inquisition (Amrán, R, 2013), so he had to reconcile with the Church in Cordoba. Likewise, the Lucena family had new problems with the Inquisition in Zaragoza, since it is known that the prothonotary was involved in the defense of a brother retained by the Holy Office (Riera, C & Riera J, 2005). Perhaps also due to his humanist education.

The humanist Ruy Lopez de Segura, a Spaniard, considered the first unofficial world chess champion, who apparently fled to Peru where he spent his last days[3], deserves special attention. In his book, 'Libro de la invención liberal y arte del juego de ajedrez', published in 1561, in its first chapter he says: "Chapter I, game of chess, game of science and mathematical invention". Although Lopez assured that the new game was created in Castile —I agree with him—, he agrees with the poets in the sense of being an invented game; that is, new and unprecedented and, also, in the sense of being a game of mathematical invention; that is, astronomical[4], since the orbits and conjunction of the

---

[2] One of the reasons why the Inquisition appeared in medieval Europe was precisely because of false conversions, particularly of Jews. In general, it is believed that the Jews continued with their Mosaic practices.

[3] It has been speculated that Ruy Lopez traveled to Peru by order of the Crown of Spain or the Church of Rome; however, there is no evidence of such an order. What is certain is that after his trip to Peru, his acquaintances in Spain never heard from him again, not even when his brother Alonso returned from Peru in 1590. In his will he does not name Ruy, perhaps because he had died or he knew that he would never return because Ruy would have commissioned him to build a convalescent hospital in his hometown.

[4] In the Middle Ages, and even in the Renaissance, pure mathematicians were not frequent. The usual thing was that mathematical developments were carried out hand in hand with astronomers. In fact, it was common for the knowledge of astronomy, geography and mathematics to be united in the same scientist (Revista Pensamiento Matemático. Vol. IX, N° 2. Set..2019). Moreover, this mathematical invention is related to the Golden Number, whose application is corroborated by

planets Mars, Venus and Mercury referred to in the poem "Scachs d' amor", follow a strictly mathematical pattern. The most serious, it seems to me, is the defiant tone when he says that chess is science and that *"science is that which can be demonstrated [...] and not that which is demonstrated with words"* (folio 2). For a church incapable of supporting its own teachings and that, in fact, contradicted its principles by setting fire to those who questioned its doctrine, the words of a cleric like Ruy Lopez constituted true blasphemies.

Lopez is considered a cleric and a humanist, and strangely forgotten in his time. Apparently, his humanist education prevented him from tolerating the actions of the Inquisition. Humanism and the Inquisition represented two diametrically opposed currents of thought in 16th century Spain. While humanism promoted education, critical thinking and tolerance, the Inquisition was associated with censorship, repression and religious orthodoxy.

It is very strange and suspicious that they overlooked the inventive nature of modern chess, especially if it is said, literally, by the poets Castellví, Vinyoles and Fenollar, whom they consider "fathers" of modern chess, repeated by the "father" of chess theory, none other than the Spaniard Ruy Lopez de Segura; as strange and suspicious as their temporary oblivion in history and almost total absence of documentation illustrating the life of Lopez after embarking to Peru. How can they make amends to the fathers of the invention?

The end of the skein

But, let's ask ourselves, why were chess and its exponents persecuted? Did the new chess contain something special that affected certain interests? What I discovered was disconcerting: the new chess may contain codes or models that reproduce on the board the harmony that the Spanish sages, many of them Jews, saw in the cosmos. At the time, certain information was sensitive for a certain organization since its disclosure questioned their legitimacy, credibility or, ultimately, their authority to present themselves as "representatives of God on Earth". At the time, the Catholic Church had adopted a literal interpretation of certain biblical passages that seemed to support a geocentric view of the universe. For example, reference was made to passages such as Joshua 10:13, which speaks of the Sun "standing still in the midst of heaven,"as evidence that the Earth was the center of the universe.

The truth is that, seen in perspective and with a modern view, the arguments of the church were self-serving falsehoods for evidently political reasons. The truth about the celestial bodies and their relationship to the Sun, however, followed a dangerous but fruitful path. The Ptolemaic view defended by the church was losing weight in the face of the contributions of the Arab astronomers Al-Battani and Al-Zarqali, known in medieval Europe as Albategnius and Azarchel, respectively, who provided accurate information about planetary orbits and the position of the Sun. Like them, the works of Georg von Peuerbach and Regiomontanus also contributed to the heliocentric theory. These 15th century astronomers and mathematicians developed improved astronomical models and tables that influenced Copernicus and helped him to perfect his theory.

However, I do not know if the distances of the planets to the Sun were known before 1543, which is when Copernicus made his theory public; according to his approach, those distances would be (Ilalux Observatory, 2020):

PLANET		DISTANCE (AU)
Mercury		0.386
Venus		0.719
Mars		1.520
	Σ	**2.625**

Now, these values that add up to 2,625 are expressed in astronomical units (AU); that is, taking as a unit the distance from the Earth to the Sun. Of course, I do not pretend to make a scientific demonstration in astronomical matters because, frankly, I am completely ignorant of its laws. Nevertheless, encouraged by the results of these calculations —which could very well be a coincidence— I would not pass up the opportunity to state my point of view. I understand that planetary orbits, stellar dynamics and gravitational interactions in the universe are explained by the laws of physics, such as Kepler's laws for planetary motion and Newton's theory of gravitation. However, in the absence of a scientific explanation that demonstrates the application of the golden number in the distances of the planets to the Sun, I come across these strange relationships precisely with the application of the enigmatic number, where Φ is the golden number and its value is approximately 1.618033988:

PLANET	FORMULA	ESTIMATED (AU)
Mercury	$(\Phi-1)^2$	0.382
Venus	$2(\Phi-1)^2$	0.764
Mars	$4(\Phi-1)^2$	1.528

These values are in almost exact correspondence with modern measurements since, for example, for Mercury the difference is (0.382-0.387) = -0.005; 1.29%. On Venus it is 0.041; 5.6%. On Mars it is verified in this way, (1.528-1.524) = 0.004; 0.26%, as can be seen in the following comparative table:

PLANET	FORMULA	ESTIMATED (AU)	REAL (AU)	DIFFEREN %

Mercury	$(\Phi-1)^2$	0.382	0.387	1.29%
Venus	$2(\Phi-1)^2$	0.764	0.723	5.67%
Mars	$4(\Phi-1)^2$	1.528	1.524	0.26%

You may wonder how I discovered this relationship between the distance of the planets and the golden number. I tell you, in my previous publication, Chess Reform and the Golden Number, I had found that the value of 2.625, which is the sum of the distances to the Sun of the planets that inspired the *Scachs d' Amor*, is a value linked to one of the properties of this fascinating golden number: $\Phi^2 = 1+\Phi$. I then hypothesized that, 2.625 = 1+1.625, assuming that 1.625 is the golden number, because the latter, at the same time, is the quotient of 1456/896, which are the total mobility values of the queen and rook in modern chess and which, pardon the repetition, are numbers that follow the Fibonacci progression. Then I used the figure of the golden rectangle with those values, a square of side 1, plus a rectangle of side 1.625, as I show in the following image:

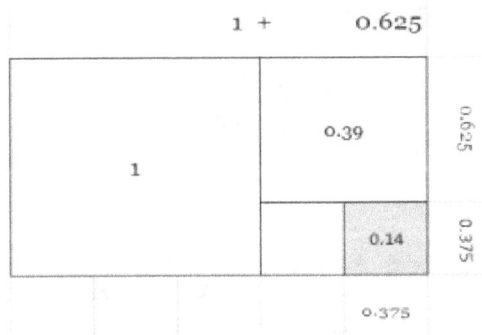

Here it can be seen that $0.625^2 = 0.39$, the distance of Mercury, but $0.625 = (\Phi-1)$. The most surprising thing was to come across the story of the "ghost planet", Vulcan, whose existence was defended even by Le Verriere, a French astronomer of the 19th century, who was right about the existence of Neptune. He estimated that the ghost planet should be at a distance from the Sun of 0.1427 AU. Knowing the properties of the golden rectangle in the sense that, whether you add or remove a square, the resulting figure is also golden, I set out to repeat the previous operation and found, to my surprise, the value of 0.14 by doing the operation of 0.375^2. Applying the golden number I obtained: = 0.1406, which in the graph is 0.375^2, very close to Le Verriere's value. This was the end of the skein. At this point I thought (because of that "transitive" property of golden rectangles) that, if it works "inside the rectangle", it will also work "outside", adding squares, so we will obtain larger golden rectangles. The rest was trial and error.

The proposal

My proposal is shown in the following table and shows values consistent with the idea that the planets have, in terms of their distances, a harmonic relationship with the golden number. These values deviate from the true values by 3.8% on average, in absolute terms; and it is better than the Titius-Bode Law which, on average, deviates by 5.99%, up to Neptune, and 15.97% if Pluto is included.

PLANET	FORMULA	ESTIMATE AU	REAL AU	DIF ABS	DIF %
Mercury	$(\Phi-1)^2$	0.382	0.387	0.005	1.29%
Venus	$2(\Phi-1)^2$	0.764	0.723	0.041	5.67%
Mars	$4(\Phi-1)^2$	1.528	1.524	0.004	0.26%
Jupiter	$[6(\Phi-1)^2]^2$	5.252	5.205	0.047	0.90%
Saturn	$[8(\Phi-1)^2]^2$	9.337	9.582	0.245	2.56%
Uranus	$[12(\Phi-1)^2]^2$	21.010	19.190	1.820	9.48%
Neptune	$[14(\Phi-1)^2]^2$	28.600	30.050	1.450	4.83%
Pluto	$[16(\Phi-1)^2]^2$	37.350	39.480	2.130	5.40%
			Average desviation		3.80%

Please look at the *Formula* column, you will see that it is an ordered progression based on the distance from Mercury. Assigning the letter M to the distance of Mercury and replacing it in the series, we have the following table:

PLANET	$M=(\Phi-1)^2$	ESTIMATE AU	REAL AU	DIF ABS	DIF %
Mercury	M	0.382	0.387	0.005	1.29%
Venus	2M	0.764	0.723	0.041	5.67%
Mars	4M	1.528	1.524	0.004	0.26%
Jupiter	$(6M)^2$	5.252	5.205	0.047	0.90%
Saturn	$(8M)^2$	9.337	9.582	0.245	2.56%
Planet X	$(10M)^2$	14.590	???		
Uranus	$(12M)^2$	21.010	19.190	1.820	9.48%
Neptune	$(14M)^2$	28.600	30.050	1.450	4.83%
Pluto	$(16M)^2$	37.350	39.480	2.130	5.40%
			Average desviation		3.80%

This is my final proposal.

Now look at the second column containing the progression, something strange happens there; you will notice that there is a factor: the number 2. The passage from one planet to another implies adding two units to the factor; thus, the formula for each planet runs two by two: 2, 4, 6, 8, 12, 14 and 16. And the number 10? Between Saturn and Uranus there should be a planet whose distance to the Sun would be: $(10M)^2 = 14.59$ AU.

The progression can be represented in the following conditional formula:

$$d_n = \begin{cases} n.M; & \text{si } 1 \leq n \leq 4 \\ [2(n-2)M]^2; & \text{si, } n \geq 5 \end{cases}$$

Where:

- d_n is the n-th planet estimated distance.
- n is the planet index.
- M is the value of the distance of Mercury which is $= (\Phi-1)^2$

And it can be proved in the following table, which includes the planet Earth in the third position, and the not yet discovered Planet X in the seventh position. It is necessary to specify that the Earth, being the unit of measurement, does not require a formula and its value is 1 which, at the same time, is the Astronomical Unit. Another important point that jumps to the sight is the double formula that coincides with the nature of the planets, rocky and non-rocky. For the first ones, such as Mercury, Venus, Earth and Mars, known as inner planets, the formula is $d_n=n.M$; for the rest of the planets —non-rocky— which are known as outer planets, $d_n=[2(n-2)M]^2$, where $n \geq 5$.

PLANET	TYPE	n	FORMULA $M=(\Phi-1)^2$	ESTIMATE AU
Mercury	Rocky planets d=n.M	1	M	0.382
Venus		2	2M	0.764
Earth		3	-	1.000
Mars		4	4M	1.528
Jupiter	Non-rocky planets $d_n=$ $[2(n-2)M]^2$	5	$[2(5-2)M]^2$	5.252
Saturn		6	$[2(6-2)M]^2$	9.337
Planet X		7	$[2(7-2)M]^2$	14.590
Uranus		8	$[2(8-2)M]^2$	21.010
Neptune		9	$[2(9-2)M]^2$	28.600
Pluto		10	$[2(10-2)M]^2$	37.350

If this proposal were to consistently validate the distances of the planets from the Sun, it could suggest the possibility of a significant discovery: the existence of a hitherto unknown planet. It would not affirm with certainty the existence of such a planet; however, mathematical consistency would support its presence. As the astrophysicist Mario Livio points out, mathematics is the language of the cosmos, and the convergence now of a mathematical pattern that speaks for nature, the Golden Number, suggests a new perspective in understanding the structure of the solar system.

Before delving into the conclusions, it is crucial to clarify that my intention does not lie in a direct comparison with modern scientific standards. Rather, I intend to highlight the astonishing skill of medieval savants, particularly the persecuted Jews who, despite lacking the advanced tools of contemporary astronomy, would have used the Golden Number and Golden Rectangle to confirm planetary distances.

This knowledge takes on an even more surprising dimension when considering the historical context: a time when astronomy was in its first steps, without radars, space telescopes or satellites orbiting the Earth. Even more amazing if we take into account that those who ventured into this knowledge risked their lives. Could it be that this persecution was related to the revelation of these astronomical secrets? And, if this is the case, in the face of imminent danger, was this knowledge encrypted in chess?

Note that this progression was not only used to predict the possible existence of the planet Vulcan, but it did so long before its non-existence was confirmed in the 19th century. This historical nuance adds a fascinating layer to my research, seeking to highlight the connection between this medieval approach and the astronomical discoveries that would come later.

Conclusions:

1. The present proposal is constructed on the basis of the value of the Mercury distance, which is expressed by the formula $M=(\Phi-1)^2$. This choice of the golden number provides an essential foundation for the sequence.
2. The estimates of the distances of the planets to the Sun, applying the golden number, deviate from the real distances by 3.8%, on average, and result in a more refined proposal than the Titius-Bode Law.
3. All planets are covered by the progression, even the imaginary planet Vulcan whose distance would have been $M^2=0.1406$ AU.
4. The progression excludes Ceres as a planet, but includes Pluto.
5. The distances of the inner planets to the Sun, except for the Earth, which is the unit of measurement, follow a geometric progression of ratio 2, whose first term is $M=(\Phi-1)^2$: AU
6. The distances of the outer planets follow a quadratic progression, where the first term is Jupiter with the value of $(6M)^2$ AU and the rest follows the formula $d_n=[2(n-2)M]^2$, for $n \geq 5$. The second order ratio of this progression is $8M^2$; which means that the distances between neighboring planets, as they move away from the Sun, will be increasing by $8M^2$ AU.
7. Planet X, which scientists are looking for beyond Neptune, would be found around $(10M)^2=14.59$ AU.
8. A planet beyond Pluto would be at a distance of $(18M)^2 = 47,271$ AU.

References

Argos1942. (August 20, 2020). *El origen judío del ajedrez moderno (The Jewish origin of modern chess)*. Obtained from MetaJaque: https://metajaque.info/?s=origen+judio

Higuera de Frutos, S. (September 15, 2019). *Historias de las matemáticas (Stories of mathematics)*. Obtained from *Revista de Pensamiento Científico (Journal of Scientific Thought)*: https://acortar.link/vM378Z

Instituto Astronómico Porta Coeli AC. (January, 2020). *Observatorio Ilalux (Ilalux Observatory)*. Obtained from: https://acortar.link/zsL94a

López de Segura, R. (1561). *Libro de la invención liberal y arte del juego del axedrez (Book of the liberal invention and art of the game of chess)*. Alcalá de Henares: Juan de Brocar.

Pérez Bosch, E. (2009). *Los valencianos del Cancionero General. Estudio de sus poesías (The Valencians of the Cancionero General. Study of their poetry)*. Universidad de Valencia.

Pérez de Arriaga, J. (s.f.). *Real Academia de la Historia (Royal Academy of History)*. Obtained from Lucena: https://dbe.rah.es/biografias/87450/lucena

Reyes La Rosa, J. (August 22, 2022). *Reforma del ajedrez y el número de oro (Chess reform and the golden number)*. Lima, Lima, Peru.

Riera, C., & Riera, J. (2005). *Biografía de Luis de Lucena (Luis de Lucena biography)* Obtained from Universidad de Valladolid: file:///C:/Users/Taxi%20Puntual/Downloads/Dialnet-BiografiaDeLuisDeLucena-2470611%20(1).pdf

Rubio Vela, A. (2020). *Los Castellví en la Baja Edad Media (The Castellví family in the late Middle Ages)*. Obtained from: https://acortar.link/EdcAar

Rueda Gallego, A. (s.f.). *Club Chaturanga de Ajedrez (Chaturanga Chess Club)*. Obtained from *Grandes olvidados. Ruy López (Forgotten giants: Ruy Lopez)*: https://www.clubchaturanga.com/grandes-olvidados-ruy-lopez/

www.ingramcontent.com/pod-product-compliance
Lightning Source LLC
Chambersburg PA
CBHW070948220526
45471CB00007B/2948